THE BRAIN
The Ultimate Thinking Machine

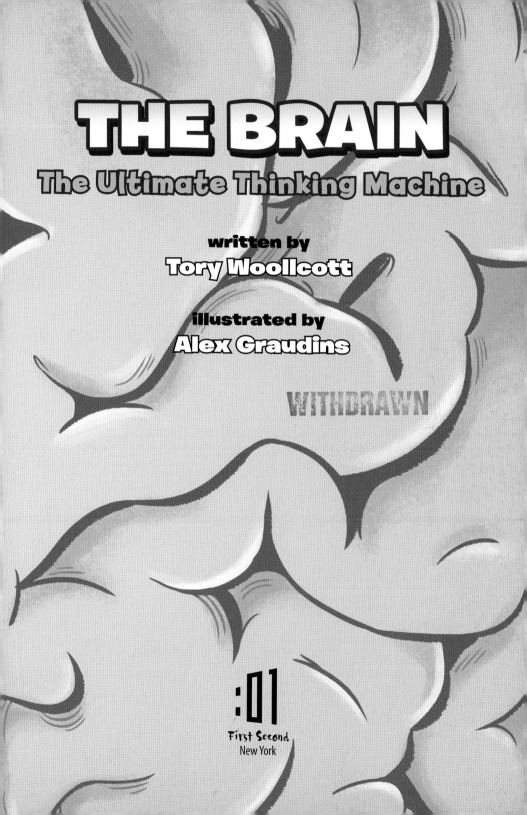

THE BRAIN
The Ultimate Thinking Machine

written by
Tory Woollcott

illustrated by
Alex Graudins

WITHDRAWN

:01

First Second
New York

For my mom and dad, who gave me this great brain.
And my granny, who showed me how to use it!
—Tory

For Mama and Papa Graudins for their endless support.
—Alex

First Second

Text copyright © 2018 by Victoria Woollcott
Art copyright © 2018 by Alex Graudins

Drawn and colored with Adobe Photoshop CS5 on a 22HD Wacom Cintiq. Lettered with Comicrazy font from Comicraft.

Published by First Second
First Second is an imprint of Roaring Brook Press,
a division of Holtzbrinck Publishing Holdings Limited Partnership
175 Fifth Avenue, New York, NY 10010
All rights reserved

Library of Congress Congress Control Number: 2017957416

Paperback ISBN: 978-1-62672-801-1
Hardcover ISBN: 978-1-62672-800-4

Our books may be purchased in bulk for promotional, educational, or business use. Please contact your local bookseller or the Macmillan Corporate and Premium Sales Department at (800) 221-7945 ext. 5442 or by e-mail at MacmillanSpecialMarkets@macmillan.com.

FIRST

EDITION

First edition, 2018
Edited by Dave Roman
Book design by John Green
Brain consultant: Jordan Poppenk

Printed in China by Toppan Leefung Printing Ltd., Dongguan City, Guangdong Province
Paperback: 10 9 8 7 6 5 4 3 2 1
Hardcover: 10 9 8 7 6 5 4 3 2 1

My love of science first began with an appreciation of the stars in the sky. And I fell in love all over again when I discovered the stars in our minds.

When I was twelve years old, my grandpa gave me a book called *Cosmos*. I loved the stories the author, Carl Sagan, shared about our universe—the billions of stars in our sky, and the millions of galaxies beyond our own. Thinking about the vastness of space made me think about our lives here on Earth, all the things we've accomplished, and what it means to be human. How did we create spaceships to take us to the moon? How did we build cities? How do we write books, and draw comics, and create art and science? How did we come up with language? How do we *think*?

Your brain is like a galaxy inside your head, with a hundred billion stars called *neurons*. Each brain is like an incredibly powerful computer. Every one of those hundred billion neurons can make up to a thousand connections with other neurons. That means there are over a hundred trillion connections in the brain—more than a thousand times as many stars as scientists think there are in our Milky Way galaxy! Like the wires sending information inside your computer, these connections send information all over your body to keep you walking and talking. These special, precious cells—and their helpers, called *glia*—are what make you, YOU! When you were born, you already had almost all of the neurons you'll ever have. They were just waiting to grow and adapt as you got older, learned new things, and became your own person.

We still know very little about our universe, and we know even less about the brain! Every day, neuroscientists, psychologists, and cognitive scientists chip away at tiny questions to try to better understand our minds. Scientists are using all kinds of incredible tools and technologies to try to figure it out— like changing the DNA of cells to make them glow in rainbow colors under a microscope, or shining lasers on neurons that have special light-activated proteins to control the brain! Still, we're nowhere *near* being able to answer all the big questions. Questions like: What is a memory? Where does personality come from? Are there any other animals with minds like ours? How can we heal the brain after an injury, or prevent the diseases that destroy it?

These unknowns were part of what made me fall in love with neuroscience. As a neuroscientist, I get to ask my own questions about the brain: How do all of our brain cells work together to help the brain grow and develop? How do the proteins made by different kinds of cells affect the ways that neurons connect and communicate with one another? How do they go wrong in diseases? Like all scientists, I hope that, someday, I'll know the answers to some of those questions—and I'll share those answers with you.

In the next pages, you'll get to meet two sisters, a mysterious mad scientist, and his zombie butler who will all work together and illustrate the things we've learned in neuroscience so far. While you read, spend some time thinking about *thinking*. Isn't it crazy that you're using your brain to read a book about the brain, learn new things about the brain, and store that information in your brain so you can retrieve it later and share it with someone else—so they can repeat the whole process with their *own* brain? Given how mind-boggling it is to even try to imagine *that*, maybe it's not a surprise that we still know so little about the brain!

For now, enjoy learning about what we *do* know—and by the time you've finished reading, scientists may have learned something new!

—Alison Caldwell, aka Alie Astrocyte,
Writer and Host of Neuro Transmissions,
a YouTube channel all about the brain,
PhD Candidate of Neuroscience, UC San Diego, CA, USA

3

The Egyptians thought so little of the brain, they didn't even keep it when they mummified their dead! They kept all the important stuff in special containers called **canopic jars.**

Imsety (Liver)

Hapy (Lungs)

Duamutef (Stomach)

Qebehsenuef (Intestines)

Brain

Contrary to popular belief, the Egyptians didn't actually pull out the brain through the nose but through the back of the head, via the foramen magnum!

SHOOMF!

But just because you have neurons doesn't mean you need a brain! When neurons started working together, **nervous systems** began to develop!

Suddenly life went ballistic! New forms of life started to appear, because cells could now work as a team!

These new types of nervous systems can be separated into two groups:

Bilateral Symmetry appears in organisms that are divided into two symmetrical halves, left and right. Humans, birds, and worms are all examples of bilateral symmetry.

Radial Symmetry appears in organisms that don't have a distinct left and right side. Sea anemones, jellyfish, and starfish are all examples of radial symmetry.

Cephalization is when a group of neurons gather in one place at one end of the body and form a bump called a ganglion. This is a huge step for evolution!

Radial symmetry led to the development of the **nerve net**, a collection of neurons spread throughout an organism's body that functions like a brain.

Fish

During the Cambrian explosion, about 530 million years ago, fish developed the first true brain and spine. A fish's central nervous system mostly controls movement and deals with the senses!

Amphibians

About 370 million years ago, in the Devonian period, lobe-finned fish evolved into early amphibians. Their brains were very simple, and had more in common with their fish ancestors than with other land animals!

KEY

- Olfactory bulb: used to detect smell
- Forebrain: controls behavior
- Midbrain: used for motor control
- Hindbrain: controls automatic body functions

Reptiles

Reptiles evolved from amphibians around 310–320 million years ago, during the Carboniferous period.

Birds

During the Triassic period, over 230 million years ago, dinosaurs evolved! Unfortunately, we don't have any dinosaur brains lying about, but we do have their descendants—birds!

22

Neurons have all the same organelles as other cells, but they have a lot more features than a regular cell!

Nucleus: where the DNA of a cell is stored

Cell body: the body of the cell, also known as the soma

Cell membrane: the outside of a cell, sort of like a skin

Myelin sheath: a fatty covering around an axon

Axon: a projection off a nerve cell that sends *action potential* (more on this in a minute!) away from the body of the cell

Dendrites: branch-like projections coming off the cell body of a neuron that **receive** signals from other cells

Nodes of Ranvier: the gaps in the myelin sheath

Axon terminals: the ends of an axon that hold neurotransmitters

Synapses: the points where a neuron connects to another cell

Axons can vary in length a lot— the longest axon in a human is about 1 meter long. Giraffes can have axons as long as 4.5 meters, and blue whales have axons up to 20 meters in length!

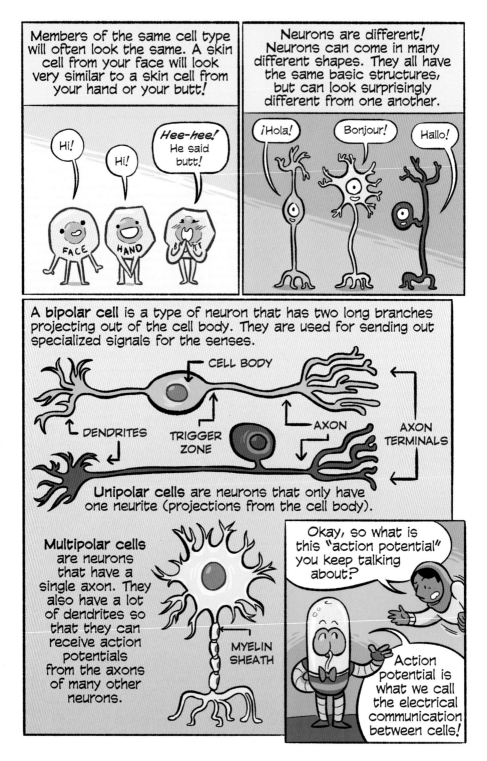

A neuron's job is to communicate with other neurons and cells throughout the rest of the body. A neuron generates electrical signals along its cell membrane and inside its cell body. These signals are called **action potential.**

1. The dendrites of the cell receive stimulus.

Yahoo!

2. That stimulus gets stored up in the cell body, or soma, of the neuron.

3. Once that cell body builds up enough of these signals, a threshold is reached, triggering the cell body to send out a signal of its own.

4. That signal is called an action potential, which then travels down the axon to the next cell.

Your brain is like a network of 100 billion batteries, and nerve signals can move at speeds of up to 540 kph (335 mph)!

You can also think of action potential as a wave of electricity that moves along the axon of the neuron!

Ion discharge down the body of the cell.

ION CHANNELS

CELL MEMBRANE

Okay, but how do neurons make action potential?

Neurons talk to each other via chemicals called neurotransmitters, which are released by axon terminals and are passed to the receptor of a neighboring cell. Each receptor can only react to one type of neurotransmitter.

Because each axon is connected to so many other neurons by its axon terminals, it's important to be picky about which neuron gets which message, otherwise things get confusing very quickly.

In effect, neurotransmitters filter out all cells the message isn't for!

Non-neuron cells found in the brain are called *glial cells*.

Hey, little glial cell!

What's new, neuron?

Unlike neurons, glial cells don't make electricity. But there are ten times more glial cells in the brain than there are neurons.

Hi! Hi! Hi! Hi! Hi! Hi! Hi! Hi! Hi! Hi! Hi!

There are four main types of glial cells:

ASTROCYTE   MICROGLIA   OLIGODENDROCYTE   SCHWANN CELLS

Each of these cells has a distinct appearance, and each has a totally different role in the nervous system!

Microglia are specialized immune cells only found in the brain! About 10–15% of all the cells in your brain are microglia.

Microglia protect your brain from intruders, and they also keep the place clean.

These cells are active! They move around the brain a lot, and they change shape in order to do this.

In order to be able to move between tightly packed nerves in the brain, microglia squish themselves very thin and pull themselves around.

The microglia become blobs when they need to absorb and remove "garbage" from the brain.

Oligodendrocytes and **Schwann cells** are pretty similar! They both wrap themselves around the axons in a process called myelination, which creates the **myelin sheath**.

NUCLEUS   AXON   CYTOPLASM

The oligodendrocytes and Schwann cells support and protect the long, delicate tendrils of the axons projecting off the nerves.

Oh no! I'm NAKED!

This is like that nightmare I had!

Myelin, the stuff that wraps around the axon, works both as insulation and lubrication.

As insulation, the myelin sheath helps to prevent the electricity of nerve impulses from "leaking" as the nerve impulses travel down the axon.

As lubrication, the myelin sheath speeds up the nerve impulses by upward of 20 times!

35

The peripheral nervous system's main job is to connect the central nervous system to the rest of the body.

There are two parts to the peripheral nervous system: the somatic nervous system and the autonomic nervous system.

The **somatic nervous system** controls the voluntary movements of your skeletal muscles—any muscles that are attached to a bone! It also signals pain, like when you stub your toe!

The **autonomic nervous system** controls all the nonvoluntary or self-regulating movements of your internal organs and glands, like your heartbeat and digestion! It also tells you stuff like if your stomach is upset!

The **sympathetic nervous system** is part of the autonomic nervous system and is designed to kick in when you're in danger.

Fight or flight: this system kicks in if you're scared or stressed out—like if you're being chased by a lion!

The **parasympathetic nervous system** is the chilling system—it brings the body back to a point of homeostasis, especially after something scared you.

40

The *central nervous system* is made up of both the brain and spinal cord, but let's start with the spinal cord! It's what connects the brain to the peripheral nervous system.

There are four main spinal nerve groups: the cervical nerves, thoracic nerves, lumbar nerves, and sacral nerves.

Each one of these gives you sensation, feeling, and movement to a particular part of your body!

Cervical nerves: These are the nerves in your neck! They're in charge of your arms, neck, and upper body.

Thoracic nerves: These are the nerves in your upper back and lower body.

Lumbar and Sacral nerves: These nerves are located in your legs and the organs below your belly button.

Sensory nerves bring information from the body to the ascending tracts.

Ascending tracts carry information to the brain, like touch and temperature but also where your limbs are in reference to your body.

Motor nerves carry commands from the brain to the muscle of the body.

Descending tracts carry information from the brain to the rest of the body. This is how you move your body and how it knows when to do its job, like breathing and pooping!

The spinal nerves (or nerve roots) are projected from the spinal cord. They exit the bones of the spine through special holes called foramen.

The spinal cord has three main functions:

1) It's how sensory information passes from the peripheral nervous system to the brain.

2) It's how commands get from the brain to the rest of the body.

3) It's a reflex center.

After the skull, the next line of defense is the blood brain barrier!

The blood brain barrier is pretty much what it sounds like—it's a border that stops bad things from entering your brain.

But there are things from the blood that your brain does need, like oxygen and nutrients, and those are allowed to cross the blood brain barrier.

However, it's still much harder to cross the blood brain barrier than it is with regular veins and arteries.

BLOOD BRAIN BARRIER

REGULAR VEINS AND ARTERIES

46

Under your skull is the meninges—
Meningitis? Isn't that a thing that makes you sick?
The ***meninges*** is the part of the body between the skull and the brain. Meningitis is actually an infection of the meninges!
There are three layers to the meninges.
The **dura mater** is a tough membrane that keeps your brain from sloshing around too much. It also keeps the **cerebrospinal fluid** from leaking out of your skull.
The **pia mater** is the layer of the meninges that is closest to your brain. The pia mater stops the **cerebrospinal fluid** from getting into the brain.
Cerebrospinal fluid? Is that, like, to cushion the brain?
Yes, exactly! Like a nice, comfy waterbed for your brain.
Cerebrospinal fluid helps stop your soft squishy brain from smashing into your skull whenever you stop or turn around!

So, *uh*, where were we? Oh yes! There are four regions of the brain. Each of these regions have distinct and important roles.

The **cerebrum** is the biggest part of your brain and makes up about 70% of it. This is where all the fun stuff happens, like thought, learning, and talking!

The **diencephalon** helps regulate body functions.

The **cerebellum** is important for movement! It tells the muscles of the body how to balance and move. This is why you don't have to consciously think about walking or reaching for something. Your cerebellum does all that work for you!

The **brain stem** helps run most of your automatic bodily functions. For example, if you're running, the brain stem helps you breathe, and makes your heart beat faster when it needs to. This is the oldest part of your brain, and it was the first to evolve!

The brain stem is made up of three main parts:

The midbrain regulates vision and hearing.

The pons help keep the rhythm of your breathing.

The brain stem regulates things you don't think about, like your heart rate and breathing, but it's also in charge of things like sneezing and barfing!

The medulla oblongata is the control center for the heart and lungs!

Oh good.

The brain stem also stimulates arousal, kicking your body into survival mode if it thinks you are in danger. A really good example of this is when you're about to fall asleep...

...and then suddenly you jump back awake for no obvious reason? That actually happens when your brain stem mixes up your body falling asleep for your body falling down!

Sometimes I get confused.

The diencephalon is made up of four main parts:

THALAMUS
HYPOTHALAMUS
PITUITARY GLAND
PINEAL GLAND

The **thalamus** is like the relay center for the brain. It collects and sorts sensory information, and then directs that information to other parts of the body.

The thalamus is also used for things like sleep, language, and learning!

It looks like an egg!

The hypothalamus regulates the autonomic nervous system. It tells you when you need to go to sleep or wake up, and whether you're hungry or thirsty. It also regulates your body temperature.

The **pituitary gland** is the "master gland" of the body, is about the size of a pea, and releases hormones that help you grow and develop. The pituitary gland also regulates your blood pressure and metabolism.

The pineal gland produces melatonin, which controls your sleep cycle.

The cerebellum helps us maintain our balance and keeps our limbs coordinated. It's also where we store all our learned motions, like martial arts techniques or how to ride a bike.

The cerebellum's job is helping you learn to use your body, but it doesn't create movements on its own. That's where the cerebrum comes in!

The **cerebral cortex** is made up of the cell bodies
of neurons. This part of the brain is sometimes called
gray matter because the cell bodies give it that color.

The axons that come off the cell bodies have connections
all over the brain. We call this axon-rich layer **white matter,**
because the fatty myelin sheaths that cover the axons are white!

The *neocortex* is 90% of the cerebral cortex.
The neocortex is like a shell that covers the entire brain.
The neocortex comprises 75% of the brain.

2.3–2.8mm

But it's
actually much
larger than it
appears.

The neocortex is made up of six layers of brain cells on
the outermost part of the cerebrum. This thin, outermost
covering of the brain is only 2.3–2.8 mm thick, but
it's actually really big when you flatten it out!

The resulting wrinkles in the brain are called the sulcus and the gyrus.

The left side of your brain is in charge of the right side of your body. All the things you hear with your right ear, and all the things you see in your right field of vision is interpreted by the left side of your brain!

The opposite is also true for the right side of your brain. It interprets everything from the left side of your body!

I thought one side of the brain was creative and the other was all about math?

That's not true! Some things like language and speech are laterally focused, but most jobs are handled by both sides of the brain.

The two sides of the brain are different, but they're just two sides of the same coin!

The two hemispheres are connected together by the *corpus callosum.* It's a thick bridge of axons that connect the two hemispheres of the brain. It's like a superhighway that lets the two sides of the brain talk to each other incredibly quickly.

Your cerebrum is divided into four lobes.
The lobes are connected to the entire brain and do a lot of different things, but their primary functions are:

Frontal lobe:
Sending out of commands and interpreting of information. The frontal lobe is also in charge of self-control.

Parietal lobe:
Interpreting of language and touch, as well as helping with spacial awareness.

Occipital lobe:
Visual integration. This is the part of your brain that lets you perceive color, size, and shape.

Temporal lobe:
The comprehension of sound and language, object perception, world knowledge, and memory.

Your personality, emotions, and ability to plan for the future is stored in your **prefrontal cortex.**
It's where what make you "you" is located.

The motor cortex controls voluntary movement of your body.

The prefrontal cortex is in charge of executive control functions.

57

The prefrontal cortex is the most connected area of the brain! This is where you're able to make decisions using previous experience as well as received stimuli.

The prefrontal cortex gives you the ability to make long-term plans, set goals, relate to other people, control your emotions, and even do things like delaying gratification!

Don't share the cake! *EAT IT ALL!!!*

ANTISOCIAL BEHAVIOR

BONK!

Would you like to share the last piece of cake with me?

One of the prefrontal cortex's most important jobs is **inhibition**. This prevents you from doing every silly thing that pops into your head!

Like kidnapping a kid and trying to steal her brain?

Yes! Exactly like that!

THOUGHTS

MEMORIES

There are two kinds of **movement**: the type you think about, and the type you don't.

I can move without thinking?

The hypothalamus regulates body temperature, thirst, hunger, and fatigue.

The medulla interprets information from the stomach and the heart.

Of course you do! All the time! Every single beat of your heart, the digestion of food—even moving your hand away from a hot surface—you don't have to think about any of that!

The type of movement that the body does on its own (except for reflexes) is part of the **autonomic nervous system**. Things like breathing, body-temperature regulation, hunger, and thirst are all part of the autonomic nervous system.

All the different parts of your body—your organs, bones, and skin—need to work together for your body to run smoothly.

There are also withdrawal reflexes!

HEY! OW!!

When your body receives a stimulus...

(in this case, my kick)

...the sensory receptors detects the stimulus and nerves then send that information up the peripheral nervous system, to the spine and central nervous system.

The action potential reaches the integration center in your spine, and that information is passed on to two different places: part of it goes to the brain...

...but motor neurons will also send directions back down the spine to the peripheral nervous system, to the original point of stimulus—causing you to react and move your leg away!

Finally, your brain will get the memo, and it will figure out what caused the stimulus. A reflex is an action your body takes without you telling it to do so. This lets you react faster and stay alive longer!

Why am I always the last to know?

The *senses* are how a living thing is able to gather information about the world around it.

Our five senses (touch, taste, smell, sight, and hearing) receive and react to three types of stimuli.

Key

= electromagnetic stimuli
= mechanical stimuli
= chemical stimuli

The senses do this by *transduction*.

That's the transformation of physical information into action potentials the brain can understand.

A lot of the brain is dedicated to interpreting and understanding what our senses tell us.

That's what **perception** is: the organization, interpretation, and contextualization of sensory stimuli.

Photoreceptors are a type of cell in the eye that absorbs light.

There are two types of photoreceptor cells: cones and rods.

Cones are how we perceive color. Most humans have three cones: red, green, and blue. Together they mix to make all the colors that we can see.

BIPOLAR CELLS

GANGLION CELLS

Rods detect light and are used for night vision. Cones require a lot of light, so when there's less light, we rely more on our rods. This is why we see less color in the dark.

Once the photoreceptors have been stimulated, they fire and activate bipolar cells, which then activate ganglion cells. The long axons of these ganglion cells are what come together to make the optic nerve.

The information travels along the optic nerve until it reaches the optic chiasm.

OPTIC CHIASM 50 mm

There, the information from the left side goes to the right hemisphere of the brain, and the information from the right side goes to the left hemisphere.

In the thalamus, the lateral geniculate nucleus processes the visual information and sends it to the occipital lobe.

WHEEEEE

OCCIPITAL LOBE

The visual information is roughly decoded by the occipital lobe and visual cortex and is sent on to the cerebral cortex to be integrated with your memories and experience.

WELCOME to the OCCIPITAL LOBE

From the visual cortex, there are two highways that take the processed visual information away to be analyzed: the dorsal stream and the ventral stream.

WOO! HOO!

DORSAL STREAM

VENTRAL STREAM

The *dorsal stream* is where the brain figures out whether something is moving, where it is, how big it is, and how fast it's moving.

The *ventral stream* takes information to the temporal lobe, where it is sorted by shape, color, and patterns, until it is recognized as familiar objects or ideas. This is where we figure out what something is!

And the nose, the mountain of the face! The crowning jewel of the head!

Smell and *taste* are senses that interpret chemicals rather than energy, distinguishing them from the other senses.

We're going to be talking a lot about snot, aren't we?

What? No! Never. I'm offended you would say such a thing.

Mucus, yes. Snot, no.

Mucus is snot.

Well, it's a very special snot called olfactory mucus. It helps you smell!

There are olfactory sensors on the roof of your nose. These receptors send little hairlike cilia into the nose to capture smell molecules.

OLFACTORY BULB

OLFACTORY RECEPTORS

OLFACTORY MUCUS

In your nose, there are 10 million olfactory receptor cilia, and there are roughly 350 types of them!

Each olfactory receptor responds to different types of **molecules.** Once the receptor is stimulated, it transmits an action potential.

The action potential then travels up the stimulated olfactory receptor until it reaches the olfactory bulb.

The olfactory bulb sends its information to several different areas of the cerebral cortex, each of which handles different aspects of smelling.

The pyriform cortex helps you recognize smells on an unconscious level.

The orbitofrontal cortex is where the axons of the olfactory sensory neurons meet with the dendrites of the brain and form little bundles, and it's where you consciously recognize smell.

The entorhinal cortex ties smell to your memories and hormones.

Hi. I'm Bud.

Papillae are the bumps on your tongue where your taste buds are!

Taste buds are little holes that suck in particles of food and detect those molecules using *microvilli*. Microvilli are like tiny hairs that can detect the different types of flavors: salty, sweet, bitter, umami, or sour.

UMAMI

SWEET

SALTY

SOUR

BITTER

The ratio of each flavor is important to our understanding of taste, and all of this is decoded in the brain.

OM NOM NOM

As you chew your food, teeny tiny pieces of the food are released and travel up your nose. Your sense of smell actually works to *enhance* your sense of taste!

What we perceive as "sound" is our brain interpreting vibrations in the air around us. These vibrations are called **sound waves**.

That *sounds* really simple—

Tee-hee-hee.

HONK!

WHHHSSH

BLAH.

BLAH. BLAH.

But seriously, though, sound gives us a lot of information that we often don't consciously realize! Pitch, loudness, direction, even language, are just a few examples.

When the ear receives a stimulus, it's your brain that changes it into hearing.

Sound is created when something vibrates! A vibration, or *frequency*, is measured in hertz—the number of vibrations per second. Human hearing spans 20–20,000 hertz. We interpret frequency as pitch.

1000 Hz

600 Hz

The height of a sound wave determines the loudness of a sound. This is measured in decibels and is all about how hard the wave hits your ear.

AMPLITUDE

TIME

The ear is a complex organ and actually has more in common with touch than it does with sight because it's a mechanical sense!

INNER EAR

SEMICIRCULAR CANALS

COCHLEA

AUDITORY NERVE

MIDDLE EAR (OSSICLES)

STAPES

INCUS

MALLEUS

EUSTACHIAN TUBE

PINNA

EAR CANAL

EARDRUM

OUTER EAR

The outer ear, or pinna, is shaped the way it is to help focus sound into the rest of the ear, like a funnel, and it helps us localize sound.

After the pinna focuses the sound in the ear, the sound passes through the **ear canal**, toward the inner ear.

As sound passes down the canal, mid-range frequencies are amplified, to help make sound and speech more clear.

The ear canal also protects the eardrum, which is delicate, you guys!

After the sound has been collected by the pinna and amplified by the auditory canal, it finally reaches the **eardrum** (also known as the tympanic membrane), causing these amplified vibrations to move into the middle ear.

WHAM

The eardrum converts the sound wave into physical vibrations that travel along the **ossicles,** a chain of three tiny bones in the middle ear.

INCUS (ANVIL)

STAPES (STIRRUP)

MALLEUS (HAMMER)

EARDRUM

COCHLEA

The vibrations from the ossicles are transmitted to the cochlea by vibrating a small membrane at the front of the cochlea called the oval window.

PUSH!

OVAL WINDOW

SPLOOSH

COCHLEAR FLUID

The cochlea is a spiral cavity filled with fluid. It's a little like a seashell deep in the inner ear. The inside of the cochlea is covered in little hairlike cells known as **stereocilia.**

OVAL WINDOW

COCHLEAR FLUIDS

STEREOCILIA

As the stapes move the **oval window,** it creates waves in the **cochlear fluid** that the stereocilia pick up.

When the stereocilia are stimulated by the movement of the cochlear fluid, they transmit action potentials along the auditory nerve.

The movement inside the cochlea stimulates nerve impulses or action potentials that then pass along the auditory nerve.

LEFT EAR

RIGHT EAR

COCHLEAR FLUIDS

YAY!

WHEE!

On the way to the auditory cortex, the sound first switches from the left ear to the right side of the brain, and from the right ear to the left side of the brain.

AUDITORY NERVE

The thalamus relays the sound information to the auditory cortex.

THALAMUS

The temporal lobe of the brain helps interpret complex sounds like language and music, as well as the basic units of sound like pitch and frequency.

In your auditory cortex, there is a map called the tonotopic map. However, it doesn't map locations—it maps auditory frequencies.

The tonotopic map decodes the information from the cochlea, and this is how we tell the difference between the sound of a car engine and your favorite song.

Hearing is a team sport! Your brain can figure out where a sound is coming from by comparing the delay between the two ears receiving a particular sound.

In 1861, Paul Broca worked with a patient who had lost the ability to form speech, even though the patient could still understand what was being said to him.

After examining the patient's brain, Broca discovered what is now known as **Broca's area.**

Wernicke's area was discovered by Carl Wernicke around 1874. He had several patients who could speak, but only gibberish, and couldn't understand anything being said to them.

Wernicke's and Broca's areas are connected by the **arcuate fasciculus.** It's like a superhighway between the two areas that allows them to communicate with each other very quickly.

Broca's area: articulation of speech and grammar

Wernicke's area: comprehension of language

The auditory association cortex allows us to interpret what we hear ("is that a bee or an oncoming train?"). It's important to be able to tell the difference.

The angular gyrus is used for reading comprehension. This area helps give meaning to things we see and hear.

A lot of your sense of hearing is dedicated to understanding the spoken word, but being able to create speech is important too.

Working/short-term memory is the whiteboard of your brain. It's where you jot down ideas but erase them quickly. Most humans can hold about seven pieces of information in their short-term memory at a time.

Long-term memory is the type of memory that lasts longer than a minute and possibly your whole life.

Sensory memory is very short-term memory that can last less than a second. These are all the tiny pieces of sensory information that keeps you aware of your surroundings.

Explicit/declarative memory is the type of long-term memory you can consciously access.

Implicit memory is when you know something without having to think about how you know it.

Episodic memory is the memory of things that have happened to you.

Semantic memory is where you store all of your facts and information.

Procedural memory is the memory of skills, or how to do something.

92

So when you're given a list of things to remember, you store that in your **working,** or **short-term, memory,** which is overseen by your frontal lobe.

You remember how to ride a bike, but you don't have to think about it because the skill is saved in your cerebellum as a **procedural memory.**

As you ride a bike down the street, the cars and the people change, so you need to be using your senses to be constantly aware of your environment.

Everything that you've passed is your sensory memory. You don't need to remember this stuff for more than a second because it only exists to give an impression of the world around you.

Clowns had their origin in ancient Greek and Roman theater, around 145 BCE.

SHUT!

The english word "clown" first appeared in 1560 and was generally understood to mean "boorish peasant."

Semantic memory is the other type of **explicit**, or **declarative**, memory. This is the type of memory where you can remember facts, figures, and information without necessarily knowing where and when you learned it.

Err, yes. So...what can I get for you?

I've forgotten!

There are a few reasons why you forget things. Either the memory failed to encode, which means the memory never passed from your working memory to your long-term memory, or sometimes the memory is actually just lost (although this is rare)!

SHORT-TERM MEMORY   LONG-TERM MEMORY

Or you might not have actually forgotten, you might have just failed to retrieve the memory. So the memory could still be there, but you're having trouble getting at it.

Your short-term memory can only hold a limited amount of information for a period of time.

If something happens that takes priority in your short-term working memory—like a clown crisis, for example—your short-term memory will switch to that instead.

So other things, like a shopping list, might be lost.

This is an example of a **retrieval cue**. That's a memory association that helps you retrieve a specific memory. It's easier to remember the first thing on a list than stuff randomly in the middle. This is known as the **serial position effect**.

When you want to move something to your long-term memory, you have to pay attention to it! The best way to do that is through practice over a period of time. That tells your hippocampus that this is important.

When you sleep, your hippocampus replays what you want to remember over and over again, firming the synaptic pathways and making the memory stronger.

You'll probably forget a shopping list pretty quickly, but that time you were in a shop and a mob of clowns came in and scared you? You might remember that for the rest of your life.

Learning is the collection of knowledge or skills over time and the ability to adapt. You can examine past experiences and apply new information to solve problems.

Learning involves strengthening and changing synapses in your brain.

The stronger a synapse is, the faster you can react to a situation using the information, skills, or behavior that you've previously learned!

Learning is a lot like a path in a forest. The first time you walk the path, it can be hard to see where you're going.

You might not even know a path is there, but the more you walk it, the more clear and obvious the path becomes.

If this path is important enough to you, it might even turn into a street or highway! That's what learning is.

If you want to keep things really simple, there are two types of non-associative learning: **habituation** and **sensitization**.

Firstly, you might see something unusual and then get used to it over time.

SKREE!

Let's say that on your walk to school, you might come across a house with a huge green elephant in the yard.

Obviously, a giant green elephant would surprise anyone, and you might be very interested, but if that same elephant is there every day, you'll get used to seeing her.

Is that a green elephant?!

Yeah, her name is Eleanor Rosewood.

The more you're exposed to the same stimulus, the less you become interested in it. This is habituation.

INTEREST

TIME

Study small: Study a little bit every day, instead of studying for hours all at once! This makes it easier for your brain to absorb and store the information. Think of it like moving 100 bricks. If you try to carry them all at once, it'll be too heavy, but carrying one brick at a time is much easier.

Sleep: Sleep, sleep, *SLEEP!* Sleep is super important in transferring knowledge from your short-term memory to your long-term memory. If you've studied small, it's *much* better to get a full night's sleep before a test than to stay up all night cramming!

So to recap (because repetition is important!): Repeat what you've learned at regular intervals, even if you think you already know it. Don't overload yourself. And get enough sleep!

That's not hard at all!

It's really not! And it'll make learning and remembering so much easier!

TYPES OF

Logic and Mathematics

A strong capacity to quantify, understand, and complete mathematical problems and to see patterns.

Linguistic

Use and understanding of language, both in speech and in thought. This is the ability to make yourself understood using language and the ability to understand language.

Spatial and Visual

The ability to visualize in three dimensions, to navigate the environment, and to reproduce mental images into a real-world space, i.e., drawing, sculpting, etc.

Kinesthetic

The ability to express one's feelings and ideas using your body. For example, the ability to score a goal in soccer or to express love and loss through dance.

There are basically eight types of intelligence.

Most people are a combination of these types, and not just exclusively one.

INTELLIGENCE

Naturalistic

The ability to understand the non-human environment, to recognize the differences between types of plants and animals, and to understand the natural world.

Musical

The ability to create, reproduce, and recognize the elements of pitch, rhythm, and tone.

Interpersonal and Social

The ability to understand the motivations of other people. To be able to interact with other people well, to be able to empathize with another person's perspective, and understand both verbal and nonverbal communication.

Intrapersonal

To be able to understand one's own goals and motivations. Being highly self-motivated and tending to think about existential questions, like "What's the meaning of life?"

(It's worth pointing out that this is just one theory of intelligence! There are all sorts of other theories out there. Everybody's brain is awesome!)

A farmer has to be able to work independently and be self-motivated. They need to understand the world around them in a literal sense, to weigh different factors like the wind, rain, and sun, and to apply these factors to different types of plants and animals with different needs.

Cartoonists need to be able to work alone and be self-motivated. They need to have an understanding of anatomy and the three-dimensional world in order to draw it. They also need to be able to use their body to help express themselves through drawing.

The mixture of different people with different types of intelligence is how we have innovation, invention, discovery, and society!

Some people have a kind of intelligence that lets them do really well in school, but there are also plenty of people who have a kind of intelligence that lets them do really well in sports, drawing, or any other number of skills! Don't ever underestimate your intelligence or the intelligence of those around you!

—GLOSSARY—

Action potential

A change in polarity across the membrane of a neuron's axon, action potential is what releases neurotransmitters, allowing neurons to talk to one another.

Axon

A projection off a nerve cell that the action potential travels down.

Bilateral symmetry

Organisms that are divided into two symmetrical halves, left and right. Humans, birds, and worms are all examples of bilateral symmetry.

Brain stem

Helps to run most automatic bodily functions (the things your body does without having to think about it), like breathing and keeping your heart beating!

Capillary

The smallest type of blood vessel. Capillaries form networks in your organs to transport oxygen and nutrients to the body's cells and to help remove waste.

Cell

The smallest unit of a living thing.

Cerebellum

Located at the back of the human brain, the cerebellum tells the muscles of the body how to balance and move.

Cerebrum

Comprising two-thirds of the human brain, the cerebrum is the largest part of the brain. The cerebrum is where decision making, imagination, and language happen!

Dendrite
Branch-like projections coming off the cell body of a neuron that receive signals from other cells.

Diencephalon
The part of the human brain that contains the thalamus, hypothalamus, pituitary gland, and pineal gland. It helps to regulate your hormones and other bodily functions.

Excitatory
To increase activity or excitement, such as in a neuron.

Excitatory neurotransmitters
Chemicals your neurons release that increase the activity of the receiving neuron.

Habituation
A type of non-associative learning where the more you're exposed to a stimulus, the less you respond to it.

Homeostasis
When your body is in a state of order and everything is normal. It's how your body likes to be most of the time—calm and relaxed.

Inhibitory
To decrease activity or prohibit chemicals from being received by a neuron.

Inhibitory neurotransmitters
Chemicals your neurons release that decrease the activity of the receiving neuron.

Instinct
A behavior in animals and people that isn't learned. For example, babies are born instinctively knowing how to suckle.

Insulation

Material that is added to an object to prevent the loss of energy. For example, insulating your home prevents the loss of heat, and insulating your axons prevents the loss of action potential!

Lateral

The side of something. In the case of an animal or a human, it often refers to the sides of an animal ("left lateral" or "right lateral").

Lobe

A clear anatomical division of an organ. In the case of the human brain, the four lobes of the cerebral cortex are the frontal lobe, the parietal lobe, the occipital lobe, and the temporal lobe.

Neurons

A cell that is specialized to send and receive nerve impulses.

Neurotransmitter

Chemicals released by neurons, and are used by brain cells to communicate with one another and the rest of the body.

Neocortex

The newest part of the brain to evolve, the neocortex is the outermost part of the brain. The neocortex is made up of six layers of brain cells on the outermost part of the cerebrum.

Nerves

A collection of axons that transmit information to and from different parts of the body.

Node

A central point where things connect to each other.

Non-associative learning
Learning from repeated exposure to stimulus.

Olfaction
The sense of smell. For example, "olfactory sensors" refers to receptors related to smell. Adapted from the Latin word *olfactus* meaning "smelled."

Organelle
A tiny specialized structure in a cell, similar to an organ in your body. It helps keep the cell functioning normally.

Perception
How you organize, interpret, and contextualize the information received from stimuli.

Radial symmetry
Appears in organisms that don't have a distinct left and right side, but rather have symmetry around a central point. Sea anemones, jellyfish, and starfish are all examples of radial symmetry.

Reflex
An action where the body reacts to stimuli without conscious thought.

Sensitization
Sensitization is a type of non-associative learning where your awareness of a particular stimulus increases over time.

Synapse
The point where two neurons meet and neurotransmitters pass between them.